3D PRINTING

and Other Industrial Tech

World Book, Inc.
180 North LaSalle Street
Suite 900
Chicago, Illinois 60601
USA

For information about other "Cool Tech" titles, as well as other World Book print and digital publications, please go to www.worldbook.com.

For information about other World Book publications, call 1-800-WORLDBK (967-5325).

For information about sales to schools and libraries, call 1-800-975-3250 (United States) or 1-800-837-5365 (Canada).

Library of Congress Cataloging-in-Publication Data for this volume has been applied for.

Cool Tech
ISBN: 978-0-7166-2429-5 (set, hc.)

3D Printing and Other Industrial Tech
ISBN: 978-0-7166-2432-5 (hc.)

Also available as:
ISBN: 978-0-7166-2449-3 (e-book)

2nd printing November 2021

STAFF

Editorial

Writer
Kris Fankhouser

Manager, New Content
Jeff De La Rosa

Manager, New Product Development
Nick Kilzer

Proofreader
Nathalie Strassheim

Manager, Contracts and Compliance
(Rights and Permissions)
Loranne K. Shields

Manager, Indexing Services
David Pofelski

Digital

Director, Digital Product Development
Erika Meller

Digital Product Manager
Jonathan Wills

Graphics and Design

Senior Designer
Don DiSante

Media Editor
Rosalia Bledsoe

Manufacturing/ Production

Manufacturing Manager
Anne Fritzinger

Production Specialist
Curley Hunter

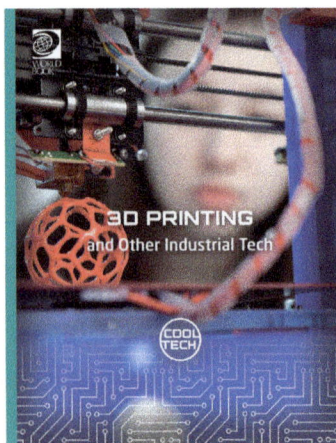

Credit: © Science Photo/Shutterstock

CONTENTS

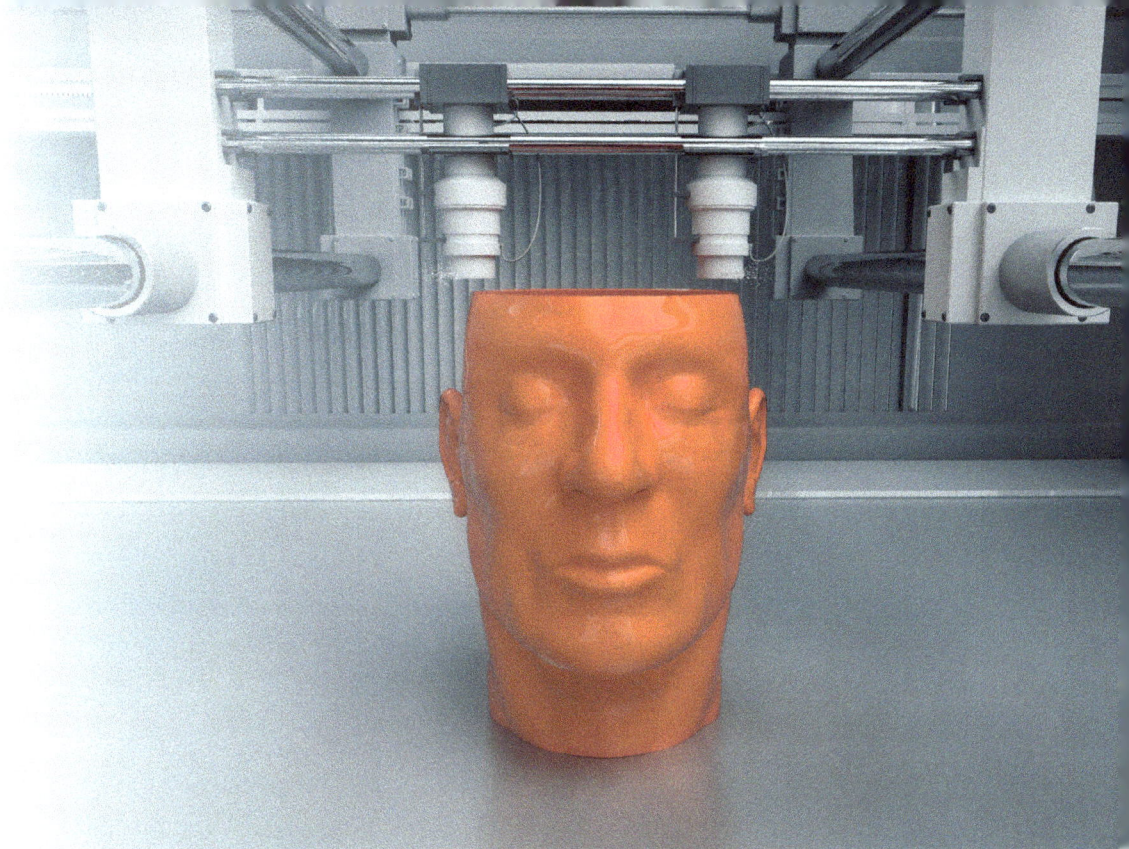

INTRODUCTION

Imagine you're building a new entertainment center for your home. You've got it nearly perfect when all of a sudden the instruction manual calls for a hexagonal wrench. Who owns one of those? And the hardware store closed an hour ago. No worries! You just tap a few commands into your computer, and your 3D printer starts whirring. In minutes, your custom-crafted hexagonal wrench is ready for use. A few quick turns and your entertainment center is ready for streaming movies and gaming in no time!

Three-dimensional (3D) printing is one of those cutting-edge inventions that is transforming modern industry to create products using the fastest, simplest, and most efficient methods. This cool new technology is making waves in many industries, including construction, electronics, manufacturing, and medicine. 3D printing devices are even available for use in your own home! Soon, you may be able to "print" your own tools, toys, and even a tasty snack!

Is 3D printing the next big thing? Or could it just be a technological fad that flames out instead of catching fire? Let's have a glimpse at some exciting new ideas 3D printing may make reality!

1 3D PRINTING

This three-dimensional printer is producing a model engine for designers to study. 3D printing makes it possible to quickly create a model or prototype, helping engineers test designs before they begin manufacturing the finished product.

THE HISTORY OF PRINTING

Three-dimensional printing is a modern twist on an ancient technology. Printing in one form or another has been around for a very long time. Over centuries, people have invented a variety of technologies to put printed words on the page.

Wood block printing began in China almost 2,000 years ago. The Chinese also invented movable **type** about 800 years later. In Europe, Johannes Gutenberg made printing from movable metal type practical for the first time around the mid-1400's. In 1875, offset printing allowed the ink text and images to be transferred or "offset" from a plate onto the printing surface. Toward the end of the 1900's, digital printing replaced traditional methods for printing books, magazines, and newspapers.

The first three-dimensional printer invented in Japan in 1981 was revolutionary—it finally lifted printing up off the flat page! Today, printing is no longer used solely for documents. Three-dimensional printers print objects!

Architects today can use 3D printers to create models of buildings. Surgeons may create 3D printed models of patients' bodies to plan surgery. Industrial designers and engineers use 3D printers to quickly generate product models for testing. Products designed may later be mass-produced by other manufacturing methods. Industries use 3D printers to make finished products too. These products include ready-made artificial limbs and specialized aircraft parts.

A NEW REVOLUTION

No matter how useful an invention may be, it cannot transform an industry or society unless it is successful. A successful technology usually means it is a technology that makes money! Inventors **patent** their creations to give them exclusive rights for the invention for a time. But if you snooze, you may lose! Then, if the invention is useful, people will buy it.

The stages of 3D printing. All 3D printing methods have three basic stages: modeling, printing, and finishing. The modeling stage uses mathematics to represent the surface of an object in three dimensions on a computer. The result is called a 3D model. The printing stage involves computer-aided design (CAD) **software.** The CAD software is used to correct any errors in the 3D model. Once corrections are made, the object is produced with a 3D printer. Finally, in the finishing stage, the surface of the printed object is made smooth and the process is complete.

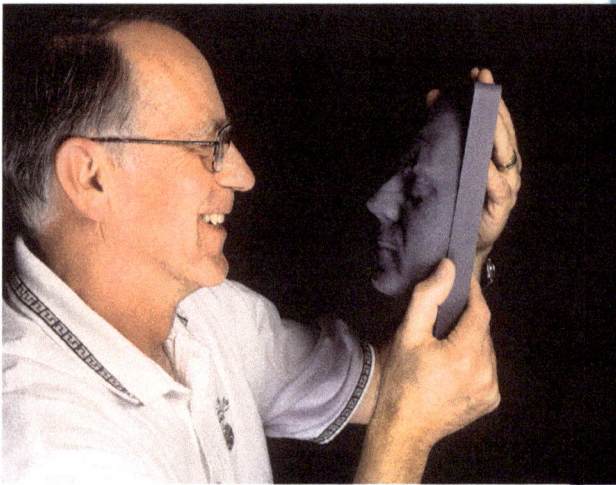

The first patent. Hideo Kodama, at the Nagoya Municipal Industrial Research Institute in Japan invented two methods for producing 3D plastic models in 1981. Unfortunately, he failed to acquire a patent for his work. A group of French engineers later investigated Kodama's ideas without any success. In 1984, American inventor Charles Hull *(above)* filed a patent for his own 3D printer, which was granted two years later.

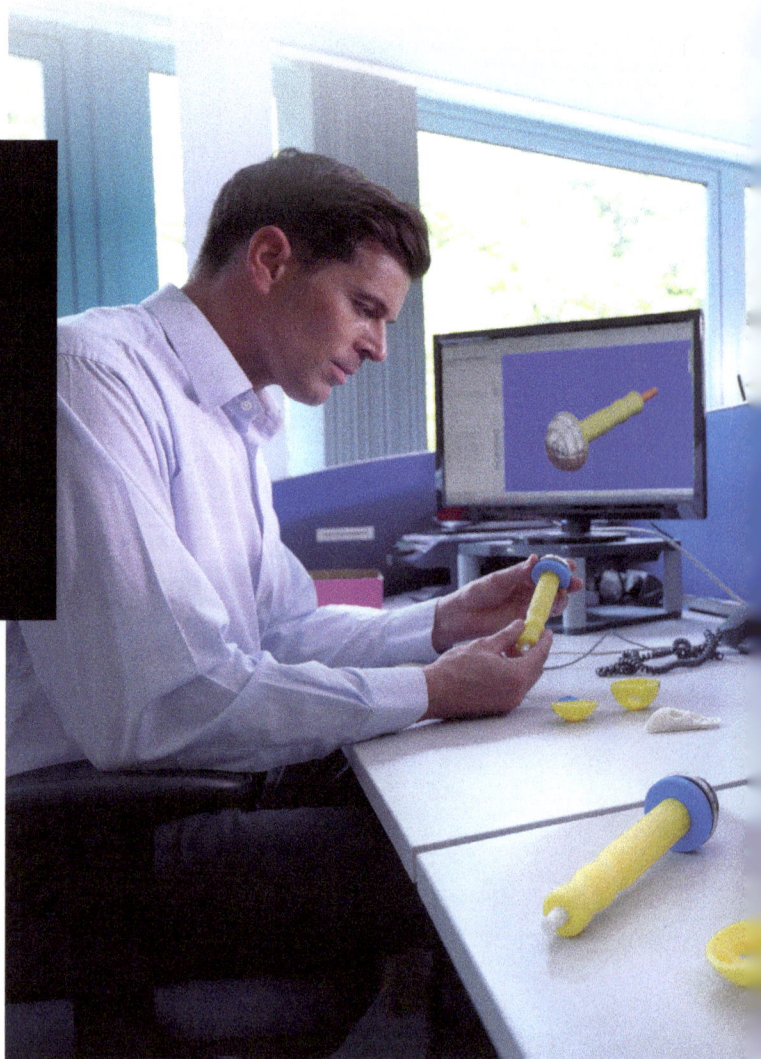

3D printing is already proving useful in a variety of applications and industries. The printers can produce objects small or large to meet almost any design and engineering need. Some 3D printed objects may even be used in the human body!

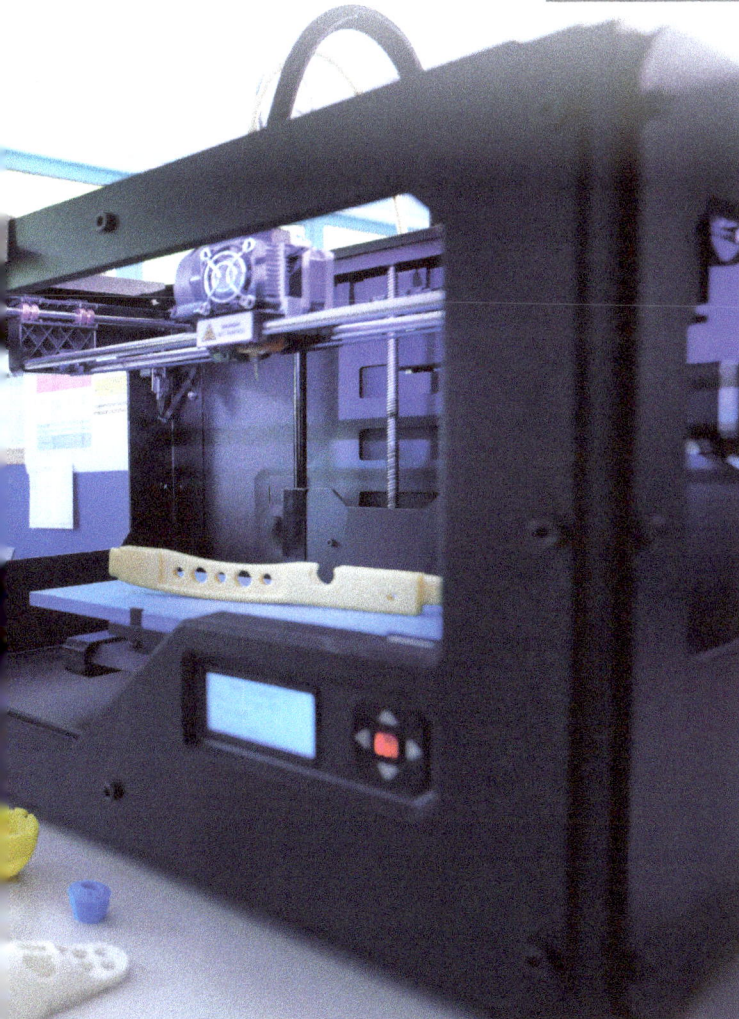

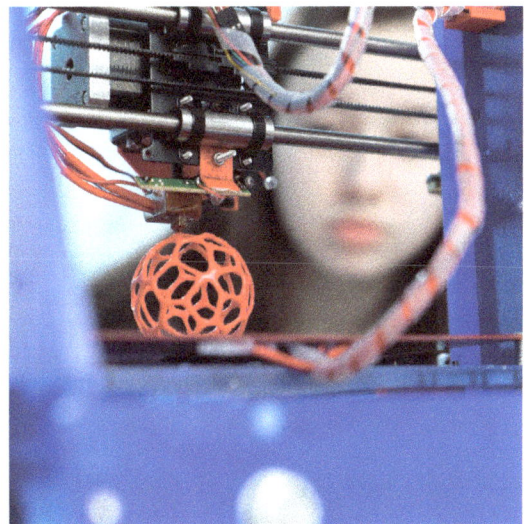

Different kinds of 3D printers. Today, there are 3D printers used by large companies on an industrial level. For example, General Electric (GE) uses 3D technology to create parts for some of the jet engine **turbines** it produces. Since 2010, the price of 3D printers has dropped, and there is a strong market for 3D printers among consumers. Today, anyone can buy a 3D printer for use in the home.

APPLICATIONS

Most anything today can be made with the help of a 3D printer. Here are just a few applications of this relatively recent technology. Who knows what innovative applications people may think of next for 3D printing!

3D food. Believe it or not, some companies are already 3D printing certain kinds of food! Three-dimensional printing is great for creating chocolate, crackers, pasta, and even pizza in novel shapes and forms. The National Aeronautics and Space Administration (NASA) is investigating 3D printing technology for making food in space.

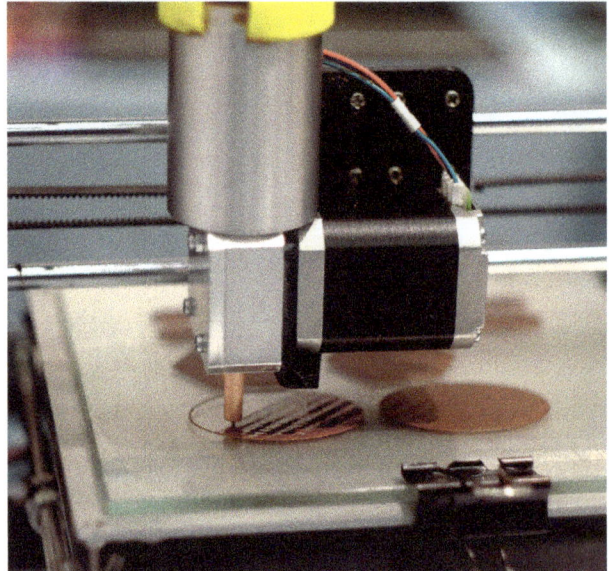

3D cars. In 2010, Urbee became the first automobile produced with the aid of 3D technology. Both the car's body and its windows were created with 3D printers. Since 2010, other cars such as the Audi RSQ have been produced with the same technology.

3D firearms. One controversial use of 3D printing is the production of firearms. In 2012, an American hardware company announced its plans to create a plastic gun that could be produced by anyone with access to a 3D printer. In 2018, a judge issued an order to prevent the company from uploading its 3D printed gun designs to the internet.

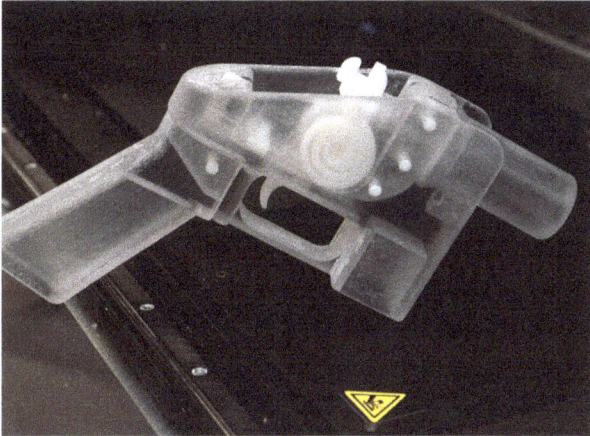

A 3D SELFIE!

Today, almost everyone takes "selfies" with their phones. Wouldn't a three-dimensional selfie be cool? With a digital camera and the right software, it's now possible to print 3D photographs of you and your friends whenever you want!

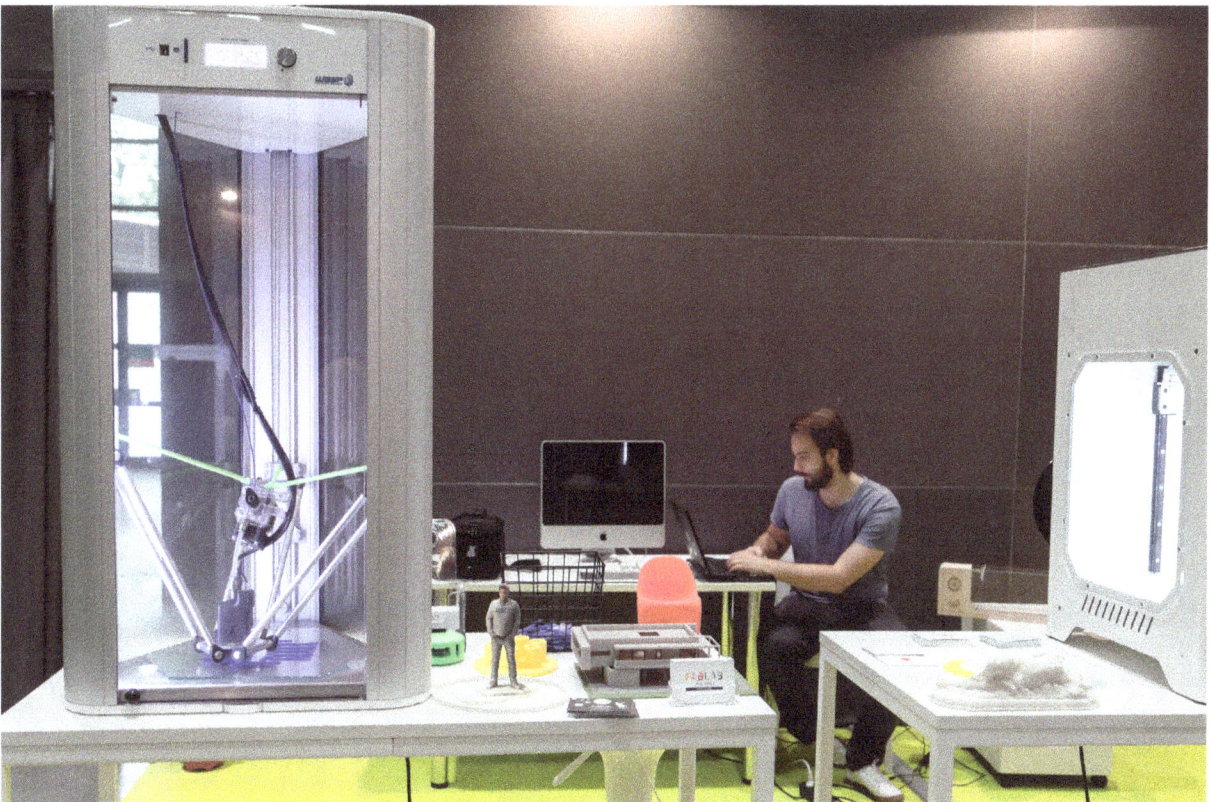

THE FUTURE OF 3D PRINTING

Cutting-edge technologies as three-dimensional printing often have unintended consequences. New inventions are not always used in the ways that the inventors imagined. As new technologies catch on, they may lead to other changes. The future of 3D printing looks amazing, but we cannot always predict the ways it may change our lives.

A changing world. Some experts have referred to 3D printing as a second Industrial Revolution. The first Industrial Revolution dominated manufacturing in the 1800's and helped bring about many social and cultural changes. Similar change may be on the horizon, thanks to such recent technological innovations as 3D printing.

Changes in medicine. The medical field has already seen incredible change from 3D printing technology. A process known as bioprinting involves the use of special materials that produce new tissues from living cells. These materials are created with 3D "bio-printers" and "bio-ink." This technology is already being used in some surgeries, but future possibilities seem limitless. There may soon come a day when doctors replace body parts—an ear, a nose, or even a heart—simply by printing up new ones!

Changes in the law. Today, many laws exist that affect how companies design and manufacture certain products. However, if 3D printing becomes as widespread as some experts predict, anyone will be able to produce these same products. New laws may need to be passed to protect existing patents, **trademarks,** and **copyrights.**

The world's first 3D-printed house *(above)* was completed in Beijing, China in 2016. Using 3D printing technology, scientists can produce custom-fitted artificial limbs and even precisely position living cells and biomaterials to "bio-print" replacement **organs** and limbs.

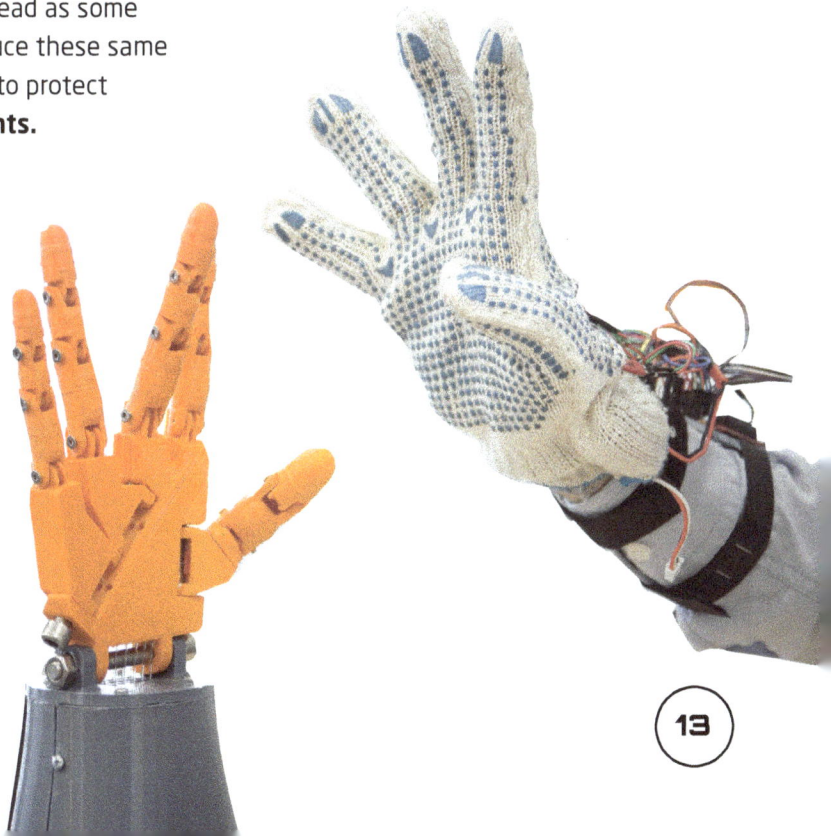

13

2 SOLAR POWER

HARNESSING THE POWER OF THE SUN

The sun is very hot, but cool technology is necessary to help make the promise of solar energy a reality! Think about all of the free energy that streams down from the sky on a bright sunny day. If we could harvest this energy, other sources that cause pollution, such as coal and other **fossil fuels,** won't be needed. Our sun is a nearly limitless source of clean energy. But capturing this energy is not easy. Sunlight is spread thinly over large areas of the Earth. Solar energy must be collected and concentrated in order to produce usable power—even after dark!

Today, solar energy is used to power anything from spacecraft and satellites to wristwatches. Solar energy is also useful in remote areas where extending electrical power lines would be difficult. But thanks to cool new technologies, the future of solar energy is so bright, you'll have to wear shades!

A GROWING INDUSTRY

Solar energy has tremendous potential to create an abundant supply of cheap, clean power. But the technological challenges that face solar energy means it is often more expensive than dirty fossil fuels. So the solar industry grew in fits and starts. Climate change caused by fossil fuels may force some hard choices. And new solar energy technologies may yet help save the day!

Renewed interest. After the energy crisis of the 1970's, the United States, Germany, and Japan were among the first nations, to prioritize solar energy technology. By 1983, the number of solar energy projects around the world had grown rapidly. But a decrease in the global price of oil slowed the growth of the solar energy industry over the next decade.

Global climate change. By the end of the 1990's, the solar industry grew as many turned to solar power as a response to global climate change caused by the widespread use of fossil fuels. **Hybrid** technology, which combines solar power with another energy source, has also become more prevalent. Examples of this technology include photovoltaic-diesel generators and photovoltaic solar collectors.

Engineers inspect photovoltaic cells in production at a factory *(above)*. Photovoltaic cells produce electricity from sunlight. This ability results from the photovoltaic effect. This occurs when the energy in sunlight causes electric charges to flow through layers of a conductive material to produce a useful electric current.

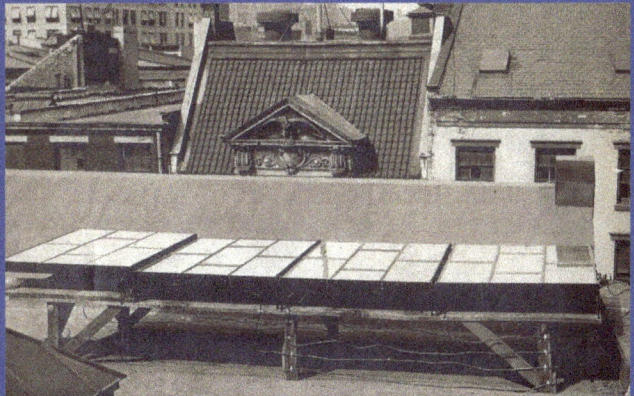

SUNLIGHT INTO ELECTRICITY

Harnessing the power of the sun dates back thousands of years, but modern solar-power technology has much more recent origins. In 1883, an American inventor named Charles Fritts built a solar cell (or photovoltaic cell) that converted sunlight directly into electricity. The following year, the first solar power device using Fritts's design was installed on a rooftop in New York City. Unfortunately, the invention converted solar energy into usable electricity at a very low rate. In 1954, scientists at Bell Telephone Laboratories invented the first photovoltaic cell that could produce a useful amount of electric power.

THE FUTURE OF SOLAR POWER

Solar energy technology is still in its infancy compared to mature technologies that utilize dirty fossils fuels or hazardous **nuclear** power. Innovative new technologies for collecting, storing, and using solar power promise to become the wave of the future. These "solar babies" may grow into energy giants!

CPVs. One promising new solar technology is the concentrator photovoltaic (CPV) system. Unlike the old photovoltaic system, CPV's use lenses and curved mirrors to focus sunlight onto small, multijunction solar cells. One example is the luminescent solar concentrator (LSC). When LSC's are used with conventional solar panels, the electrical power production increases dramatically.

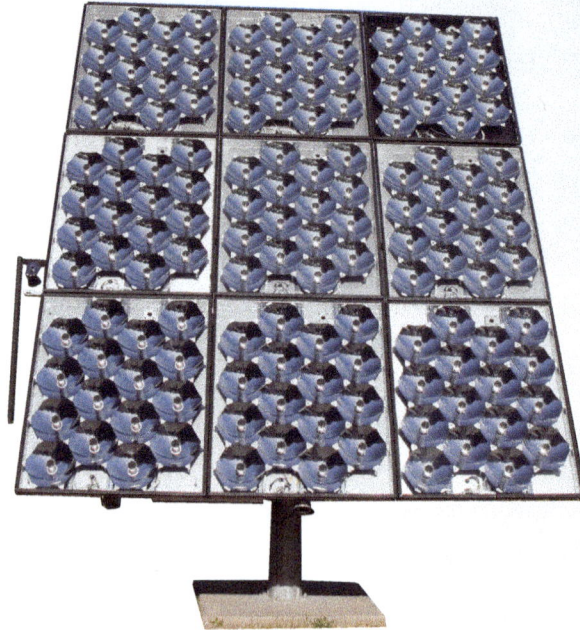

Solar trackers are a tracking mechanism that follows the sun as it moves across the sky. Trackers allow solar panels and other solar energy collectors to catch the "direct-beam" sunlight as opposed to "diffused" sunlight, which contains much of the solar energy.

Floatovoltaics are a type of solar panel used where there is limited availability of land. They are often placed in irrigation canals or water reservoirs. Typically, floatovoltaics have a higher energy conversion rate because the surrounding water keeps their solar panels at a cooler temperature than those used on land.

Solar parks. Today, solar "parks" (also known as "solar farms") already exist. Every year, more are built—and built on a larger scale. Currently, the largest one in the world is the Longyangxia Dam Solar Park in China. In 2014, the International Energy Agency predicted that solar power could produce up to 27 percent of all energy used in the world by the year 2050.

3 FUSION POWER

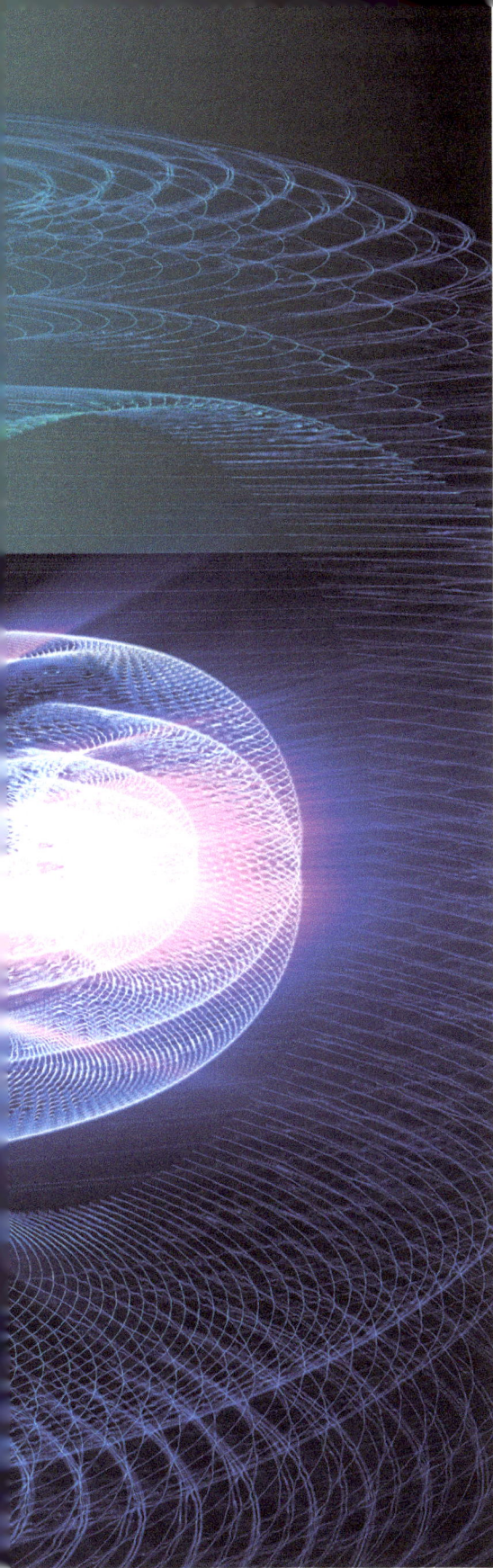

STAR POWER ON EARTH

Few things are hotter today than cool new energy technology. And among energy technologies, nothing is hotter than fusion. Not even the sun!

Fusion produces the energy that powers the sun and other stars in the sky. Nuclear fusion technology has the potential to create the power of the sun down here on Earth. Modern day nuclear plants generate power by fission—that is, splitting atoms. Nuclear fusion is a process in which two or more hydrogen nuclei—the small, dense centers of hydrogen atoms—are combined or fused together. This process releases a tremendous amount of energy.

The technological challenges involved in developing nuclear fusion into usable power are some of the most difficult problems humans have ever had to solve. Nuclear fusion only occurs when hydrogen nuclei are subjected to incredible pressures and heat to create a superhot form of matter called **plasma.** So far, creating these conditions requires more electrical power than is produced by the fusion reaction!

Nuclear fusion creates plasma so hot that scientists have yet to create a container that can hold it for more than a few seconds without melting. Other challenges involve transforming the heat generated from fusion into usable forms of energy, such as electricity.

But these problems are worth solving! Ordinary seawater provides an almost limitless supply of the hydrogen fuel necessary for nuclear fusion power. The rewards of successful nuclear fusion technology include a virtually limitless supply of pollution-free energy for all.

HOW A TOKAMAK WORKS

Tokamak is one of the leading designs today for a nuclear fusion reactor to generate power. This design was originally created by scientists in the Soviet Union (modern-day Russia) years ago. The word tokamak comes from Russian words meaning toroidal (doughnut-shaped) chamber and magnetic coil.

A tokamak uses magnetic confinement to solve the challenge of containing superhot plasma that drives the fusion reaction. A magnetic field is generated to push plasma away from the tokamak walls so it cannot melt. A strong electric current through the plasma also helps confine the hot matter to the center of the tokamak.

Inside a tokamak, hydrogen fuel is heated to form a superhot plasma. Powerful magnets are used to confine the plasma in a ring-shaped torus within the walls of the machine. The plasma particles heat up as they collide to reach amazing temperatures. Eventually, the particles overcome their natural repulsion to fuse, releasing huge amounts of energy.

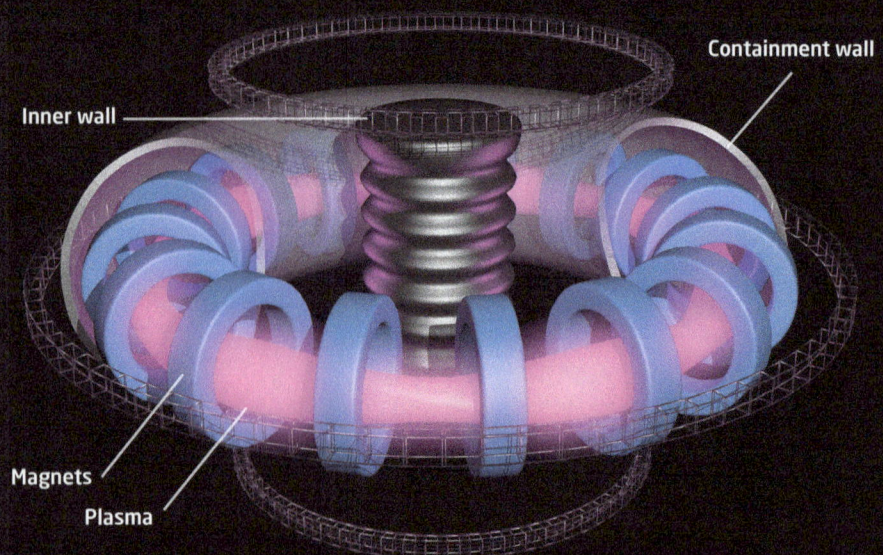

Containment wall

Inner wall

Magnets

Plasma

The first tokamak fusion reactor *(left)* was designed and built by scientists in the Soviet Union (now Russia) in the late 1950's. With their success in achieving the temperatures and confinement necessary for fusion, this design became the focus of most future research on fusion energy.

THE FUTURE OF NUCLEAR POWER

Scientists working with tokamaks have made great progress solving some of the technological challenges that stand in the way of our fusion-powered future. But they haven't put all their fusion eggs in one tokamak-shaped basket. Other fusion reactor designs may succeed where tokamaks fail. Lessons learned from these designs may also help make fusion power a reality much sooner.

The National Ignition Facility (NIF) in California is leading the way in fusion research beyond the traditional tokamak design. Scientists at NIF are aiming 192 powerful lasers to blast a gold pellet filled with hydrogen fuel to "ignite" a fusion reaction that releases great amounts of energy. By focusing NIF's laser beams onto such a tiny target, the scientists will create temperatures of more than 180 million Fahrenheit degrees (100 million Celsius degrees) and pressures that exceed 100 billion times Earth's atmosphere.

ITER. The International Thermonuclear Experimental Reactor (ITER) is the largest nuclear fusion tokamak in the world. Most experts predict this design *(seen as a model, above)* will be the first nuclear fusion reactor that generates more power than it uses to create and confine plasma.

Fusion rockets. Fusion power may take humanity to the stars. Scientists know that space exploration requires the enormous power of fusion. Perhaps a fusion rocket will one day be part of a manned mission to Mars and beyond!

NEW FUSION DESIGNS

Most research on fusion focuses on the standard tokamak design to transform room-temperature gas into the superhot plasma that fuels the drive's fusion reactions. But there are other ways to generate plasma. Worldwide, scientists are working on innovative, nonstandard fusion **reactor** designs that are smaller and more compact than a tokamak.

Scientists at the Max Planck Institute for Plasma Physics (IPP) in Germany are working with a device called a stellarator to produce fusion reactions. A stellarator is similar to a tokamak in that both designs use powerful magnets to suspend superhot plasma at the temperatures and pressures necessary to ignite the fusion reaction.

The IPP reactor, called the Wendelstein 7-X, uses 50 superconducting magnet coils to trap the plasma in a twisting and spiraling shape, rather than the doughnut shape of a tokamak. Fusion researchers have found that the twisting plasma is more stable compared to the plasma created in a tokamak. In 2018, the

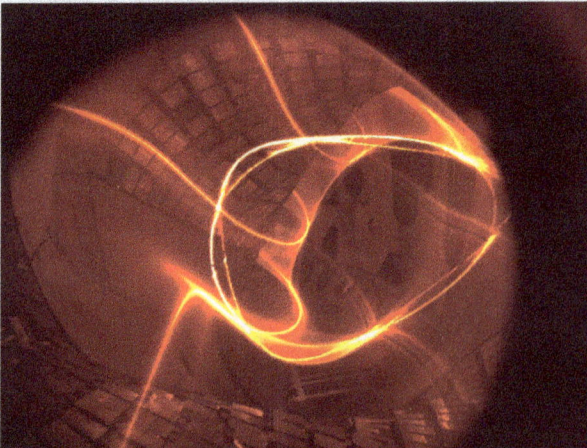

Wendelstein 7-X stellarator produced plasmas lasting up to 26 seconds. **Graphite** tiles lining the walls of the reactor allowed the reactor to achieve higher temperatures than ever before—about 40 million degrees Fahrenheit (22.2 million degree Celsius)!

At the U.S. Department of Energy Princeton Plasma Physics Laboratory (PPPL) in New Jersey, scientists are studying an innovative fusion design called transient coaxial helical injection (CHI) that eliminates the magnets in a tokamak reactor. The scientists expect the spherical fusion space will be more compact and efficient compared to the doughnut-shaped plasma produced in standard tokamaks.

University of Washington scientists originally developed their innovative dynomak fusion reactor as a class project! This reactor design, also known as a spheromak, generates magnetic fields to contain hot plasma by driving electrical currents into the plasma itself without bulky external magnets and coils found in the tokamak desgins. This allows for a much smaller fusion reactor that can generate more energy than ITER, at a fraction of the cost.

New fusion reactor designs attempt to do away with the bulky magnets while still achieving the incredible temperatures needed to start a fusion reaction within the superhot plasma. Developing materials that can withstand the searing temperatures for more than a few seconds is another challenge facing fusion research.

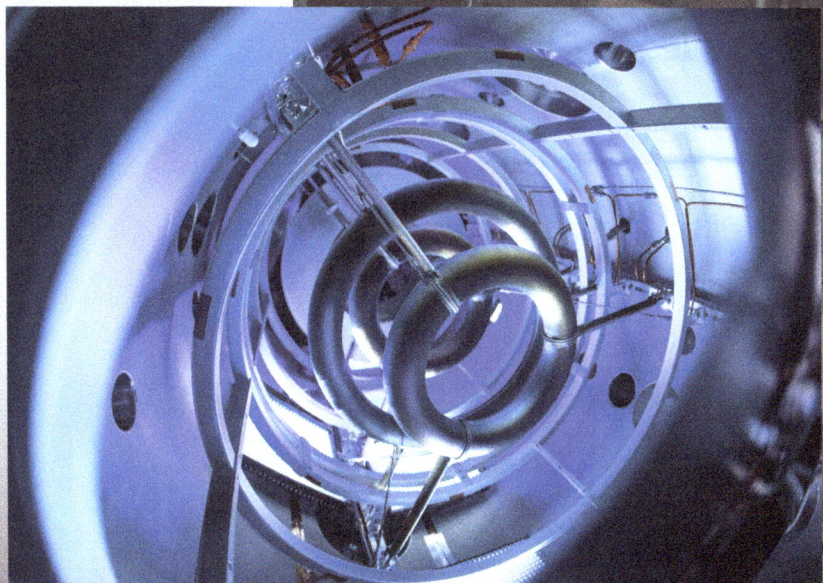

4 INDUSTRIAL ROBOTS

ROBOT TECH

Is any there technology cooler than robots? A robot is a machine that can perform an action or a series of actions automatically. Some people already own **automated** vacuum cleaners or robotic toys. In the future, people are certain to have even more robots in homes doing many different jobs.

Today, robots are a vital part of the manufacturing industry. These robots are programmed and then perform a wide variety of tasks, such as painting, labeling, and packaging. In factories, robots perform such repetitive tasks as welding and drilling. Robots are increasingly used in places that are too dangerous or impossible for humans to reach. Scientists have used robots to explore the sea floor on Earth and the surface of Mars. Robots can perform precise tasks no person can do by hand. Doctors even use robots in delicate surgery.

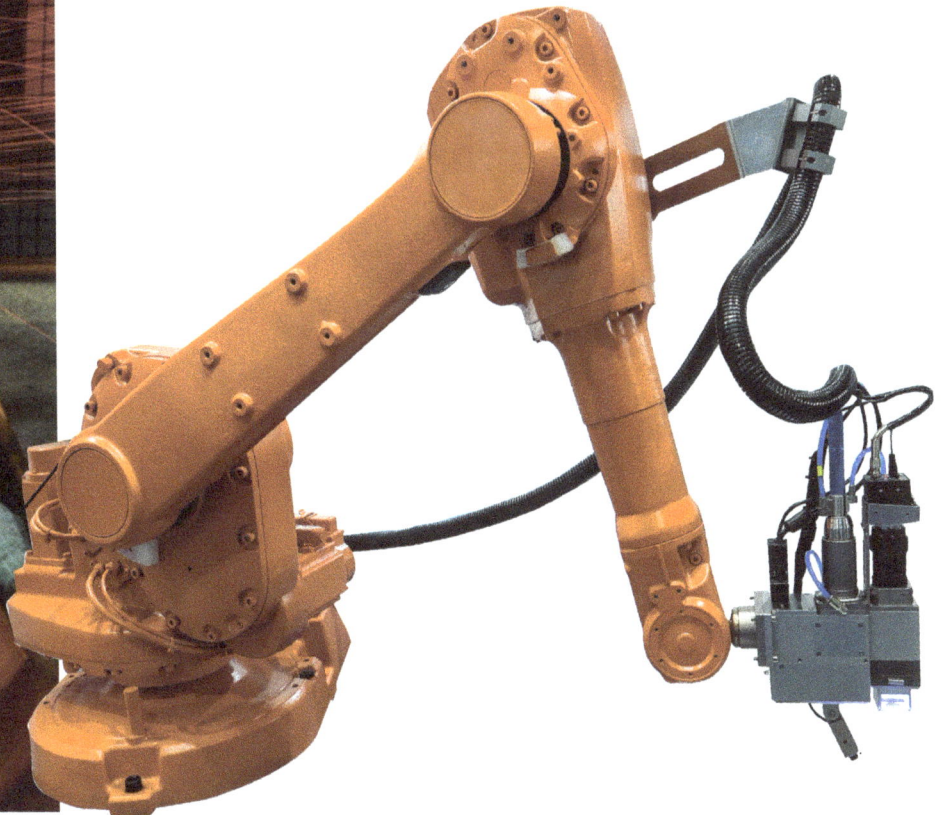

DIFFERENT TYPES OF INDUSTRIAL ROBOTS

The demands of modern industry mean that you have to have the right robot for the right job. Instead of building a single robot that can do many things, scientists and engineers are increasingly designing specialized robots for specific jobs. These new robots can do whatever job necessary faster and with more precision than any human.

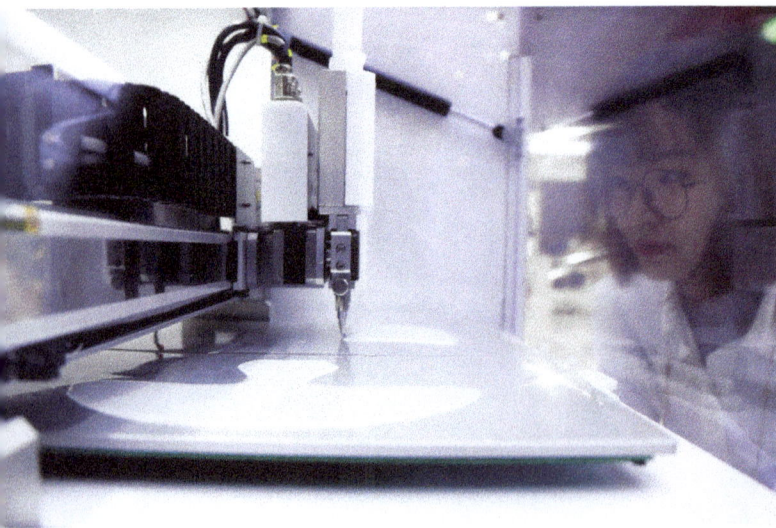

Articulated robots have **rotary** joints that allow them a wide range of movements. Simple models have two joints. More complex robots can have ten or more depending on the kind of task they were designed to accomplish.

SCARA. The Selective Compliance Articulated Robot Arm (SCARA) is a type of robot that is typically used on factory assembly lines. These robots can accomplish repetitive tasks, such as welding joints, with amazing speed.

Cartesian robots have three main joints that move in a straight line instead of rotating and move at right angles. The large cranes that are sometimes used in shipbuilding are an example of the Cartesian robots. The name comes from the French philosopher, mathematician, and scientist René Descartes.

Delta. This is a spiderlike robot that is capable of more delicate and precise movements. They are often used in the electronics, food, and pharmaceutical industries.

TELEVISION DEBUT

Unimate, the first industrial robot, proved it could perform a wide variety of tasks. Appearing on "The Tonight Show" in 1966, Unimate played golf, served beer, and played the accordion.

THE FUTURE OF INDUSTRIAL ROBOTS

For people in the workplace today, the future is robotic. Innovative robot technology will ensure that human workers and robots will be interacting in all sorts of new ways. Soon, you may not have a co-worker at your job, but a co-robot. Thank you very much, Mr. Roboto!

Collaborative robots. A collaborative robot, also known as a "cobot," is one that is designed to interact with human beings in the same workspace. Many experts in industrial robotics predict that cobots are the wave of the future. Today, human employees and cobots work together to perform quality inspection of their products. However, cobots that work alongside human beings might one day replace them. Security, for example, is a job that might become the exclusive domain of robots in the future.

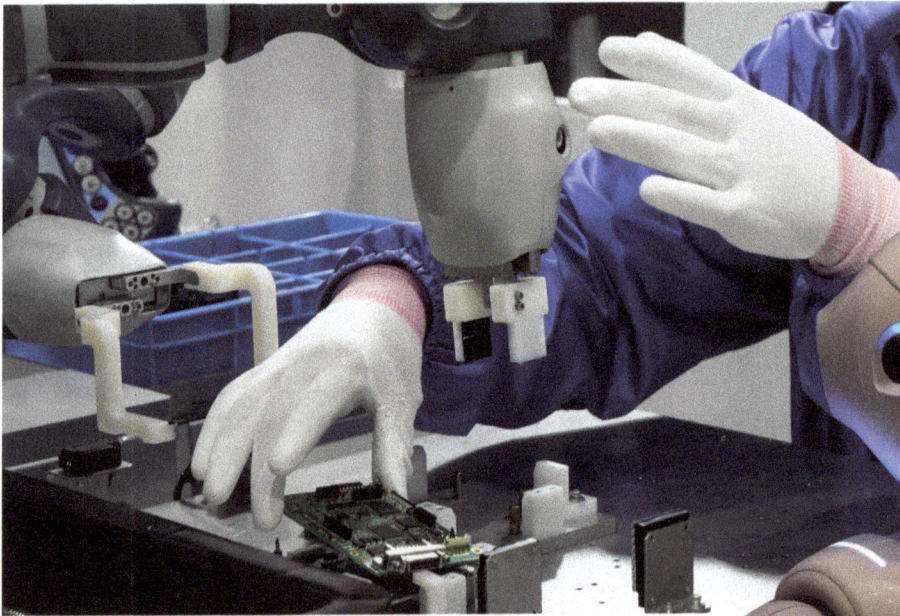

Artificial intelligence, also known as AI, will give robots the ability to perform tasks that usually require human intelligence. AI is necessary for robots to fully enter the human workforce. Steven Spielberg explored this idea in his 2001 film *A.I. Artificial Intelligence*. With new robot technology being developed today, things that were once science fiction may become a reality sooner than you think!

FANUC. The Fuji Automatic Numerical Control Corporation, also known as FANUC, is the world largest producer of industrial robots. In 2015, it released its first collaborative robot. Since then, it has worked toward developing a robot that can perform tasks not through programming but by learning through trial and error. This breakthrough will revolutionize the field of robotics as we know it.

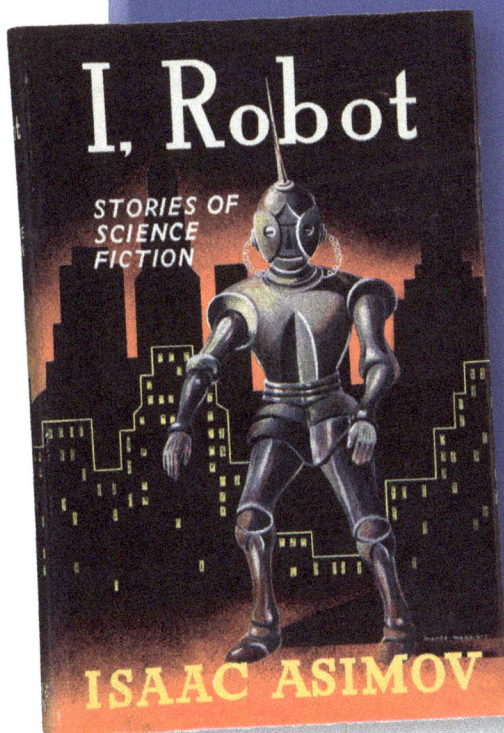

ROBOT LAW

In his classic science fiction novel *I, Robot*, author Isaac Asimov outlined three laws that he saw as necessary in order for advanced robots to assist humans. The laws state: (1) A robot may not injure a human being or, through inaction, allow a human being to come to harm. (2) A robot must obey orders given it by human beings except where such orders would conflict with the First Law. (3) A robot must protect its own existence as long as such protection does not conflict with the First or Second Law.

Asimov's work is fiction. But today, as robots are increasingly common in our lives, it may be time to develop some real robot laws!

5 NANOTECHNOLOGY

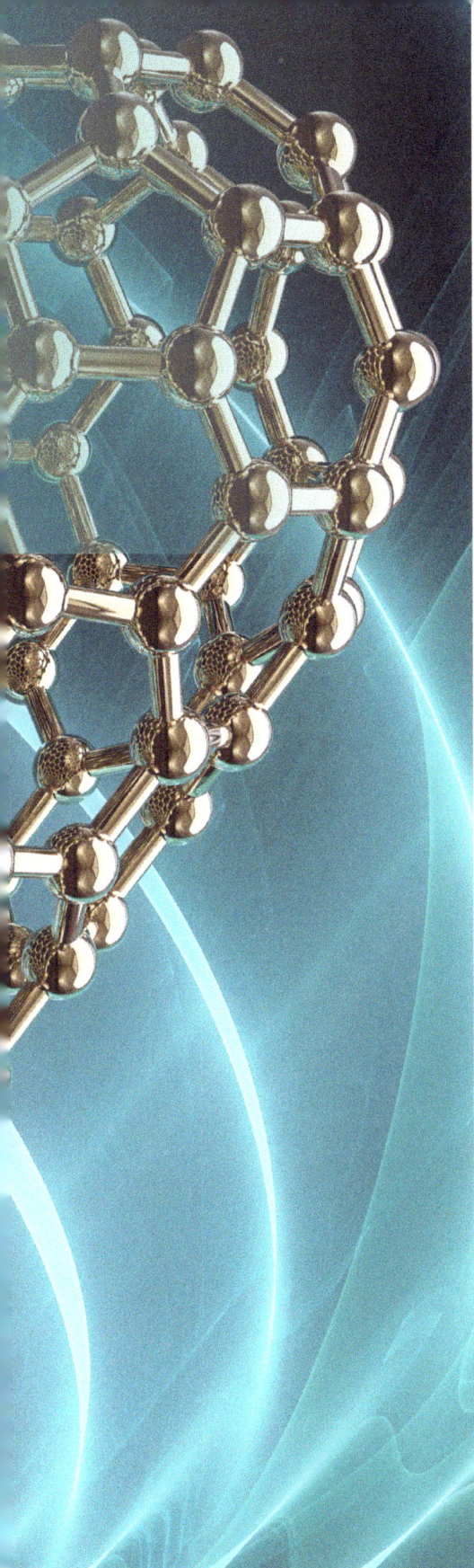

LET'S GET SMALL

In cool new technology, the next big thing is gonna be small—real small. Nanotechnology is the science of the small. In simplest terms, nanotechnology (or "nanotech") involves the manipulation of matter on an atomic and molecular level. In the future, nanotechnology may involve anything from a particle so small that it cannot be seen by the naked eye to the food you eat and the clothes you wear.

Nanotech materials, and the objects made from them, display fundamentally different properties and behavior than the same materials at larger scales. For example, tiny nanotubes (tubular structures made of carbon molecules) are about 100 times stronger than steel. They can also conduct electricity.

In the future, nanodevices may be built with these nanotubes. The devices may even include nanoscale machines that can act independently—even inside our bodies!

This tiny tech promises to deliver big changes to almost every aspect of our society.

NANOTECHNOLOGY DEFINED

In nanotechnology, objects are measured in nanometers. Nano means billionth. A nanometer is 0.000000001 meter—approximately 1/100,000 the width of a human hair or 3 to 5 times the diameter of a single atom. The term nanotechnology involves any object measured on the nanoscale (between about 1 and 100 nanometers).

NANOMATERIALS AND CURRENT RESEARCH

One day, nanotechnology may enable humans to recreate most any object in unimaginably small dimensions. In our nanofuture, computer chips may be smaller than a grain of sand. Nonoparticles may make most any fabric stronger and virtually stainproof. Medicines can be deliver directly to targeted tissues and cells, eliminating the side effects of many medications and therapies. But first, scientists and engineers must figure out how to manipulate and control nanomaterials.

Nanomaterials. A nanomaterial is any material created through nanotechnology. These include carbon nanotubes, metal rubber used as flooring material in hospitals or to cover the outside walls of industrial plants, and quantum dots, which can be embedded in living cells for medical research.

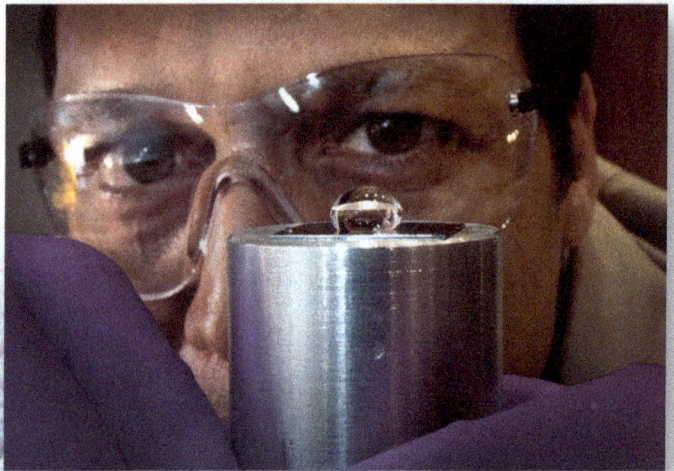

Four main approaches. In nanotechnology research today, there are four main approaches: bottom-up, top-down, functional, and biometric. In a bottom-up approach, nanotechnology uses small components to create more complex ones (this is often used in DNA research, for example). In a top-down approach, just the opposite is done: larger nanotech devices are used to create smaller ones. In the third approach, the focus is "functionality." That is, the nanotech device achieves its goal regardless of how it was constructed (synthetic molecular motors are created this way). Finally, in a biometric approach, the natural world is studied and used to design nanotech devices (this is also known as biomimicry or bionics).

Nanotech dead end? Scientists have had many ideas about how best to use nanotechnology. However, not all of them are good or work as expected. For example, silver nanoparticles, because of their antibacterial characteristics, have been tested to prevent odors in socks. But the secret to an "anti-stink" sock has yet to be discovered.

THE FUTURE OF NANOTECHNOLOGY

Many people have big dreams about how nanotechnology may transform our future. They see it as a transformative technology that will dramatically change society, health care, and many other aspects of our lives. So far, many of the ideas about nanotechnology remain dreams. But progress is being made in nanotech most every day. The promising reality of nonotechnology may not be that far off!

A nanotechnician manipulates microscopic drops of silicon to create a microelectromechanical systems device (MEMS). These devices take advantage of nanotechnology to create electronic devices and machines of incredibly small size.

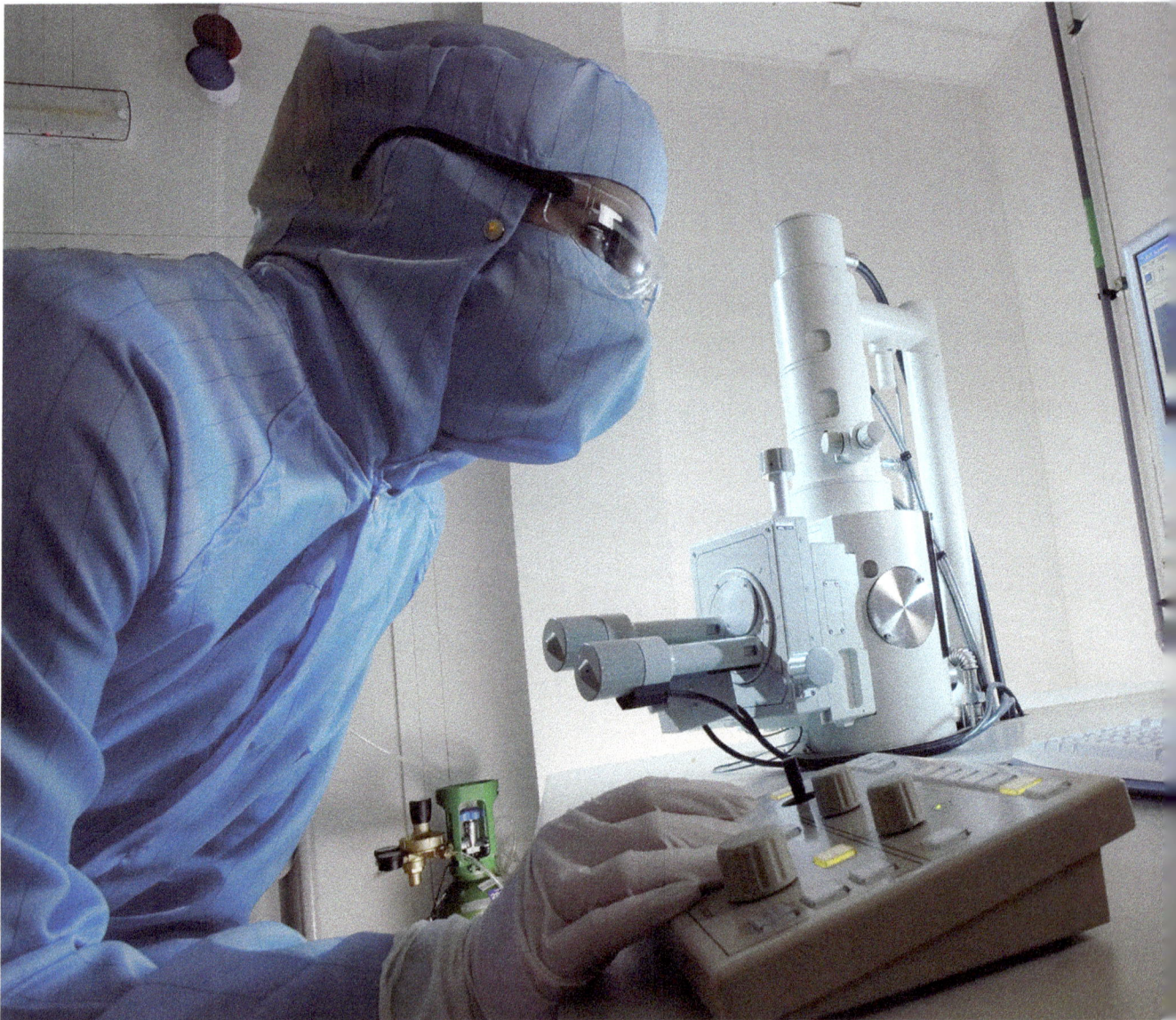

Nanorobots. The field of nanorobotics combines robotics and nanotechnology. This involves building incredibly small, yet independently acting robots. Today, the possibilities for nanorobots seem limitless. Nanobots are also known as nanomachines, nanomites, nanites, and nanoids.

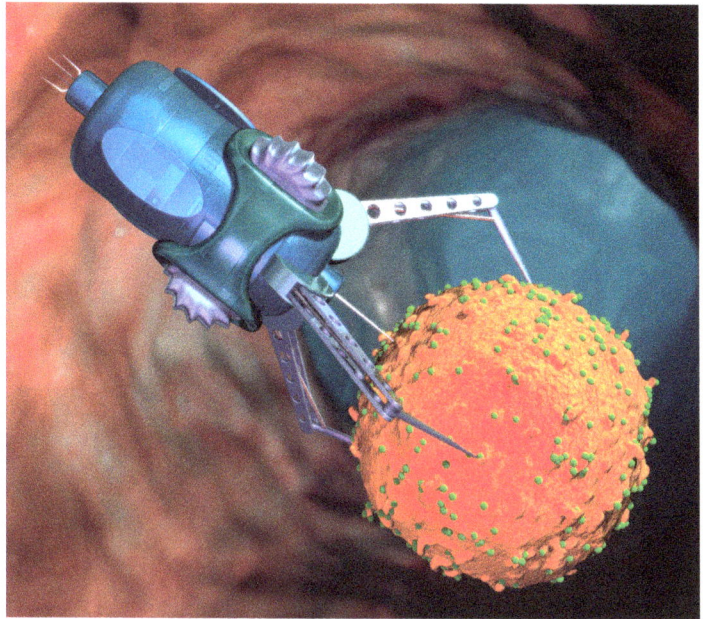

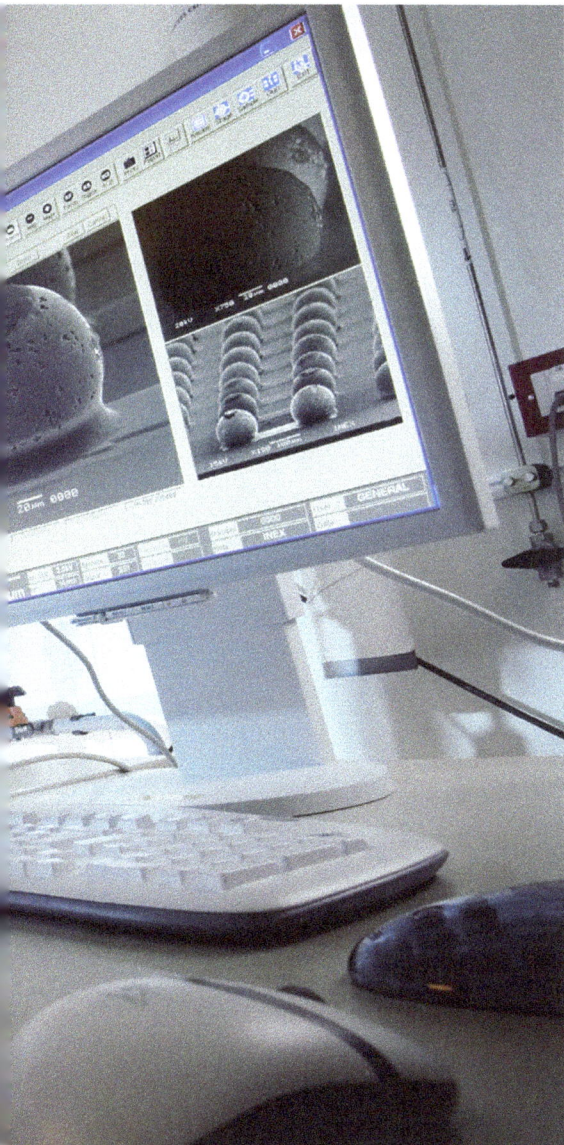

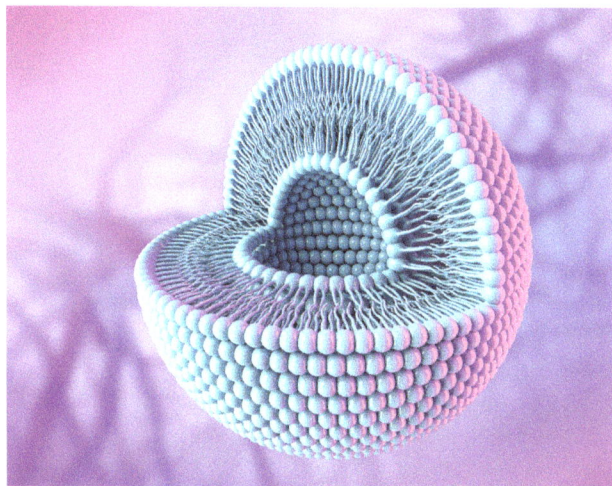

Nanomedicine. Nanotechnology may soon revolutionize medicine. Medical injections with nanoparticles may one day replace **chemotherapy** as a treatment for cancer and other diseases. Nanotechnology is also being investigated for use in certain types of surgeries, blood purification, and diabetes monitoring. Doctors may one day use nanobots to repair damaged nerves at the cellular level to restore movement and function.

6 THE TECHNOLOGY OF TOMORROW

FUTURE TECH

The future promises to be full of cool new technology that can scarcely be imagined today, except perhaps in science fiction. How these technologies will transform life and society remains to be seen. But some kinds of technology being developed today provide clues to what our technological future has in store. These cool technologies include graphene and smart fabrics.

GRAPHENE

Graphene is material made up of a single layer of carbon atoms. If three million sheets of graphene were stacked on top of each other, it would only be about one millimeter thick. Despite its small scale, graphene is the strongest material known. It has several useful properties as well, such as being a good conductor of heat and electricity.

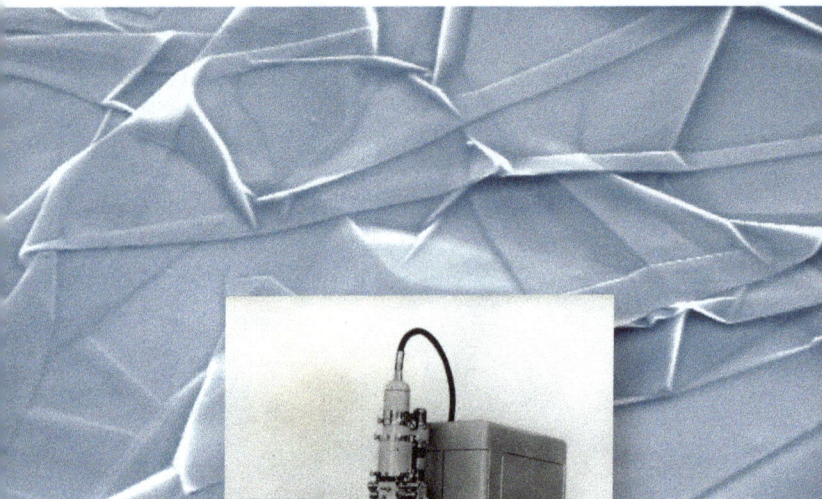

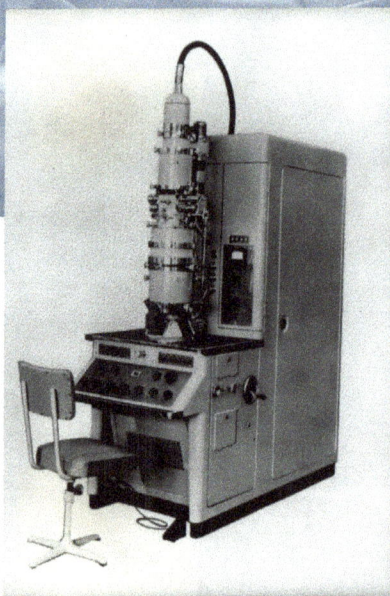

Geim and Novoselov. In 2004, two physicists named Andre Geim and Konstantin Novoselov studied graphene and learned a great deal more about it. For their study of graphene and its potential, Geim and Novoselov were awarded the Nobel Prize in Physics in 2010.

The discovery of graphene. While scientists theorized about graphene for many years, it was not discovered until 1962. Viewed through a powerful type of microscope known as an electron microscope, graphene's honeycomb-like structure was revealed. At the time, however, it could only be studied when it supported on metal surfaces.

Possible uses. Graphene has a great potential for use in many fields, including medicine, electronics, and energy. In the last decade, scientists and engineers have done research into ways that graphene might one day be used. These include removing salt from ocean water, cleaning up radioactive waste, and fully charging a **smartphone** in less than ten seconds!

Graphene has the potential to change technology and society the same way that plastics changed our lives in the past century. The first graphene materials are just now beginning to have an impact. The future seems unlimited!

SMART FABRIC

Your next set of new clothes may not just look smart, they may be smart! Smart fabric is cloth or material that has digital or electronic components embedded in it. These components can be a light, a battery, or even a small computer. Also called e-textiles, these smart fabrics are dazzling on the fashion show runway!

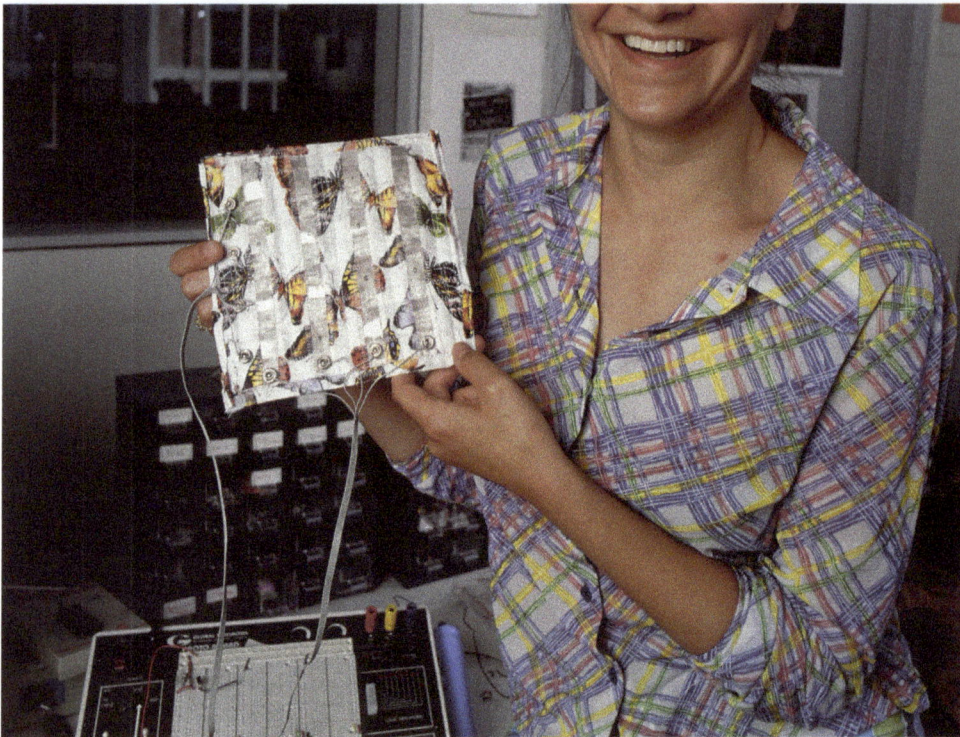

Classifications. There are three main groups into which smart fabric is divided: one in which digital or electronic sensors are attached to a garment (for example, wearing a fitness tracker), one in which the sensor is embedded in the garment (for example, a light-emitting diode of LED), and one in which the sensor and the garment are one and the same (this is known as "third-generation" smart fabric).

Some smart fabrics can sense and react to environmental conditions—think a shirt that keeps you cool when the temperature rises.

Or it may change color to suit the mood. A smart fabric biking suit may sense wind reistance and even measure the wearer's heart rate to enhance performance. A smart shirt can track information on heart rate and movement. The data may be sent to a doctor or a computer app that measures stress levels.

Smart fabrics in military uniforms may automatically sense and provide warnings when dangerous gases and other hazards are present. They may help identify friendly troops and distinguish them from the enemy.

Designer Maggie Orth *(left and below)* holds a swatch of her smart fabric that contains alternating strips of conducting and non-conducting material. This fabric can be integrated into an outfit to serve as a "keyboard" for a wearable computer!

SMART SHIRT

In 2016, the fashion design company Ralph Lauren revealed its first "smart shirt." Attached to the shirt were sensors that tracked such biological statistics as heart rate, number of steps taken, and energy exertion.

Wearable computers. Smartwatches and other devices allow us to wear a computer and take it with us anywhere. However, wearables such as the smartwatch are distinct from e-textiles. Smart fabric is not always wearable and is sometimes used in interior design, for example. Smart wallpaper and drapes may help keep your home comfortable and energy efficient.

GLOSSARY

automated moving or acting by itself.

chemotherapy the treatment of disease and infection by means of drugs and other chemicals.

copyright the exclusive right to publish or sell and otherwise control an original work that can be reproduced by printing.

fossil fuel an energy-providing material formed from the long-dead remains of living things. Fossil fuels include coal, natural gas, and petroleum.

graphite a soft, black form of carbon with a metallic luster.

hybrid combining two or more functions or modes of operation.

nuclear of or having to do with atomic energy.

organ any part of an animal or plant that is composed of various tissues organized to do certain things in life. The eyes, stomach, heart, and lungs are organs of the body.

patent a government grant which gives a person or company sole rights to make, use, or sell a new invention for a certain number of years.

plasma in physics, is a form of matter composed of electrically charged particles. The sun and the other stars consist of plasma.

reactor a device for producing atomic energy without causing an explosion.

rotary having parts that rotate.

software the instructions and routines required for the operation of a computer or other automatic machine.

smartphone a portable telephone equipped to perform additional functions beyond calling, such as providing internet access, supporting text messaging, or taking photographs.

trademark a mark or symbol owned and used by a manufacturer to distinguish their goods from the goods of others.

turbine an engine or motor in which a wheel with vanes is made to revolve by the force of water, steam, or air.

type a piece of metal or wood having on its upper surface a raised letter, figure, or other character, for use in printing.

INDEX

ACKNOWLEDGMENTS

5 © Dabarti CGI/Shutterstock

6-7 © Stock DD Video/Shutterstock; Public Domain

8-9 © Evan Hurd, Alamy Images; © Monty Rakusen, Getty Images; © Richard Wareham Fotografie/Alamy Images; © Science Photo/Shutterstock

10-11 © Maksym Kaharlyk, Shutterstock; © VCG/Getty Images; © Keith Beaty, Toronto Star/Getty Images; © Pau Barrena, AFP/Getty Images; © Tinxi/Shutterstock

12-13 © BSIP/UIG/Getty Images; © VCG/Getty Images; © Photo Oz/Shutterstock

14-15 © BrightSource Energy; © RelaxFoto.de/Getty Images

16-17 © Maximilian Stock Ltd./Getty Images; © Ruslan Dashinsky, Getty Images; © Michael Reinhard, Getty Images; Public Domain

18-19 © Ellen Isaacs, Alamy Images; © Mr. Fotos/Shutterstock; © Galyna Andrushko, Shutterstock; © API Guide/Shutterstock; © Chee Hoong Loh, Shutterstock

20-21 © Dani3315/Shutterstock

22-23 U.S. Department of Energy; © VRX/Shutterstock; © Thierry Campion, Gamma-Rapho/Getty Images; © Oleg Kuzmin, TASS/Getty Images

24-25 © Eric Feferberg, AFP/Getty Images; © Patrick Landmann, Science Photo Library; © VCG/Getty Images; NASA

26-27 © Stefan Sauer, picture alliance/Getty Images; © Matthias Otte, IPP; © Lockheed Martin

28-29 © Maximilian Stock Ltd./Getty Images; © Asharkyu/Shutterstock

30-31 © Agencja Fotograficzna Caro/Alamy Images; © Tang Yanjun, China News Service/Getty Images; © loonger/Getty Images; © Westend61/Getty Images; © Gamma-Keystone/Getty Images

32-33 © Zapp 2 Photo/Shutterstock; © Mandel Ngan, AFP/Getty Images; © Friso Gentsch, picture alliance/Getty Images

34-35 © Laguna Design/Science Photo Library; © Andrey Prokhorov, Getty Images

36-37 Brookhaven National Laboratory; © Giro Science/Shutterstock; © Dino Fracchia, Alamy Images; © Kateryna Kon, Shutterstock

38-39 © Colin Cuthbert, Science Photo Library; © Roger Harris, Getty Images; © Maurizio De Angelis, Science Photo Library

40-41 © Rost9/Shutterstock; © Monty Rakusen, Getty Images

42-43 © Andre Geim, Konstantin Novoselov, The University of Manchester; © Marcel Antonisse, AFP/Getty Images; © Zuma Press/Alamy Images; © Monty Rakusen, Getty Images; © Robert Brook, Science Photo Library/Getty Images

44-45 © Sam Ogden, Science Photo Library; © Ralph Lauren; © Nate Meepian, Shutterstock

www.ingramcontent.com/pod-product-compliance
Lightning Source LLC
Chambersburg PA
CBHW042108210326
41519CB00064B/7590